Ajay Singh, Anil Ghimire, Niroj Dahal

Energy Saving at Kathmandu University. A Project Design to Lower Energy Consumption and Costs in the Girls' Hostel, Block-17

GRIN Publishing

Bibliographic information published by the German National Library:

The German National Library lists this publication in the National Bibliography; detailed bibliographic data are available on the Internet at http://dnb.dnb.de .

Imprint:

Print and binding: Books on Demand GmbH, Norderstedt Germany
ISBN: 9783668553545

This book at GRIN:

http://www.grin.com/en/e-book/376770/energy-saving-at-kathmandu-university-a-project-design-to-lower-energy

KATHMANDU UNIVERSITY

SCHOOL OF ENGINEERING

DEPARTMENT OF MECHANICAL ENGINEERING

A REPORT ON

"PRELIMINARY ENERGY AUDIT OF BLOCK – 17 (GIRLS' HOSTEL) OF KATHMANDU UNIVERSITY"

PROJECT MEMBER

AJAY SINGH

ANIL GHIMIRE

NIROJ DAHAL

August 8, 2015

EXECUTIVE SUMMARY

This project report draws attention to the energy management process and opportunities in girls' hostel (Block-17) of Kathmandu University and different energy saving strategy with economic analysis and risks involved.

The major energy consumption in building is the lighting sources and consumes 73kWh/day in average with maximum loading of 8.652 kW. The major lighting sources, which consumes electricity, are Fluorescent Tube Lights (FTLs) and Compact Tube Lights (CFLs) used in building lighting design. The calculated cost of electrical energy by these sources are NRs.22746.30 under certain assumptions.

The expected saving opportunities in building are by meeting state-of-the-art in lighting technology achieved by replacement of CFLs and FTLs by LED technology and Boost the building by green energy. The first opportunity save electricity 39kWh/day in average and thus reduction in cost by 53.3%. The second one option goes to green energy but have higher cost of energy. The both saving opportunities have its own financial and environmental risks involved.

The two model is developed in above discussed opportunities; i) Replacement of CFLS and FTLs by LED in stages and ii) Solar-Grid Hybrid System. The each developed model save energy and also its cost and both have low environmental and financial risks involved in it.

In conclusion of this report, the synergism of both model makes one more appealing than the oters for implementation decision in girls' hostel which not only save the energy but also the cost of energy with reduction of maximum demand of energy

Table of Contents

LIST OF FIGURES

LIST OF TABLES

1. INTRODUCTION

1.1 Background

Energy use in buildings currently account for about 32% of the global total final energy consumption in the world [1]. Electric lighting is a major energy consumer in typical residential buildings. Enormous energy savings are possible using energy efficient equipment, effective controls, and careful design. The residential buildings in the urban centers of Nepal mainly use electricity and LPG for lighting, heating and cooling purposes [2].

Electric lighting design also strongly affects visual performance and visual comfort by aiming to maintain adequate and appropriate illumination while controlling reflection and glare. Lighting is not just a high priority when considering residential building design; it is also a high- return, low-risk investment. By installing new lighting technologies to meet the state-of-the-art, one can reduce the amount of electricity consumed and energy costs associated with lighting.

This project mainly concentrates on the energy saving potentials in lighting of Girls' Hostel Block (Block -17), covers total area of 2604.00 sq. meter, of Kathmandu University located at Dhulikhel [3].

1.2 Objective

The objective of this Preliminary Energy Audit is primarily to assess current lightning scheme and the viability of implementing an energy efficiency upgrade of the facility including a subsequent Investment Grade Audit.

This objective will be achieved by:

- Identifying a suitable energy performance indicator for existing and target energy use to quantify the potential for energy savings.
- Identifying a suite of measures, including savings and implementation budget,
- Identify any potential technical, financial risks to the project as currently defined.

1.3 Scope of Assessment

The assessment includes the electrical energy consumption in lighting the building with possible saving opportunities by replacement of energy efficient lighting technologies and uses of renewable energy sources and the financial assessment.

The assessment has certain assumptions which are:

- The operating hour of each lighting sources are same for all days.
- There is no any seasonal load variation.
- The survey data is same for all days.
- The cost of energy (kWh) is taken to be NRs. 9.3/kWh, which is weighted average cost of energy for non-commercial tariff rate, and demand charge is NRs. 220/KVA.
- It is not considered that energy saving can be obtained by changing lighting design scheme. i.e. the scheme of lighting design is perfect.
- The light energy (lux) reaching at the table or place, used for study, is in range and the lower values are due to the ageing of luminaire. The lux measured are in ANNEX 1.

2. ENERGY CONSUMPTION

The building has total 112 rooms; 11 in ground floor, 28 in each 1st floor, 2nd floor and 3rd floor and 17 in top floor. The total CFLs and FTLs connected in building are shown in table 1 and individual details are is ANNEX 2.

Name	Numbers
CFL (1*11W)	452
FTL (1*40W)	92

Table 1: CFL and FTL Connected in Buildings

The total capacity and energy consumption of lighting load are **8652W** and **73kWh/day** respectively. The per month energy consumption cost for the above load is calculated, assuming power factor 0.8, as

= 73 kWh/day * NRs. 9.3/kWh * 30 + (8.652/0.8) KVA*NRs. 220/KVA

= NRs. 22746.30 per month.

The annual bill of electricity of building of F/Y 2070/71 shown below:

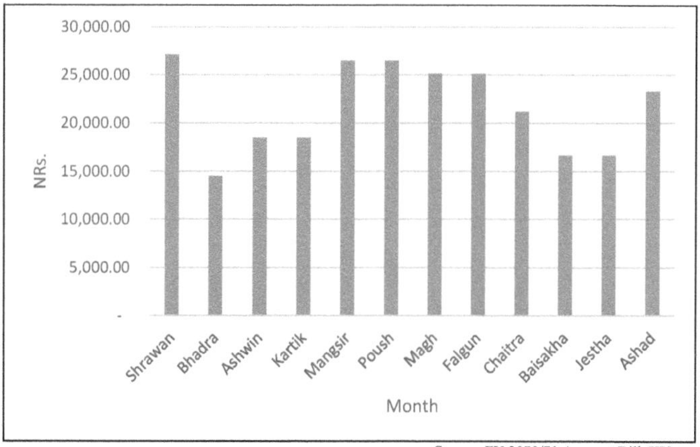

Source: FY 2070/71 Account Bill, KU

Figure 1: Monthly Electricity Bill of Girls' Hostel

The major source of energy consumption in building is the lighting source and the average energy cost for FY 2070/71 is NRs. 21628.50 per month. This figure is nearly same as the calculated value, therefore, the assumptions made are précised to the correct value.

3. EXPECTED SAVING AND ECONOMIC ANALYSIS

3.1 Expected Saving When Replaced by LED light.

The lighting design scheme is not considered to be changed and there is adequate light at day time which uses natural lighting. The expected saving can be brought by replacing the luminaire with high efficient and lower energy consumption. For this purpose, the CFLs and FTLs are replaced by LED light.

Existing Luminaire	Replacement LED light
CFL (11W)	LED (7W)
FTL (40W)	LED (12W)

Table 2: Replacement Opportunities Options

When the CFLs and FTLs are replaced by LED lights then the load and energy consumption are **4268W** and **34.3 kWh/day**. LED lights have greater power factor than the FTLs and is to be minimum 0.9. Thus the total energy cost per month is:

= 34.3 kWh/day * NRs. 9.3/kWh * 30 + (4.268/0.9) KVA*NRs. 220/KVA

= NRs. 10613 per month.

The cost saving is NRs 12133.3 per month. The comparative savings in energy, demand and cost are shown in figure 2.

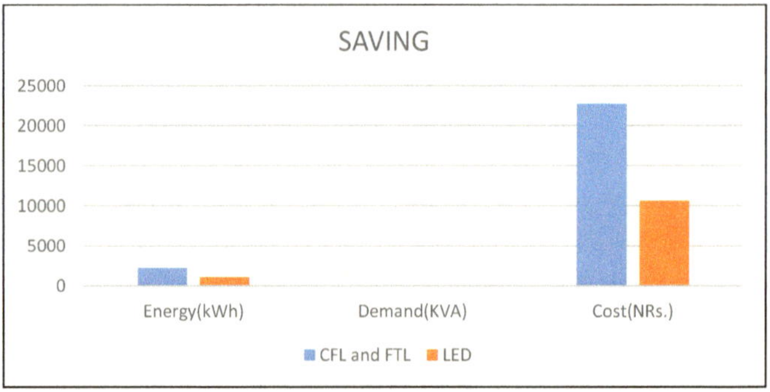

Figure 2: Comparison of energy Consumption, Demand and Cost of Energy of CFL and LED

The detailed calculations for the replacement of LED light is in ANNEX 3. From the figure 3, it is seen that the cost saving after LED light replacement is 53.3%.

The average life of FTLs and CFLs are 8000-15000 hours and LED light have life 25000-50000 hours [4] [5]. This also benefits the use of LED light.

The life cycle cost (LCC) of FTL and LED is compared with assumptions that the operating hour of light is 11 hours in a day and the average life of FTL and LED are 8000 hours and 25000 hours respectively. The energy consumption by a single FTL and LED of 40W and 12W are 0.44 kWh/day and 0.132 kWh/day. Energy bill is paid monthly. The discount rate is assumed to be 10%. When using these both for specified time as above, the life of FTL is finished in 2 years and LED light life is finished in 6 years and 3 months. Thus, the comparison is made for 6 years and the capital rates are taken from the quotation ANNEX 4.

	FTL		LED	
Time	Capital	Present Cost	Capital	Present Cost
0	120	120.00	750.00	750
1	-		-	
2	120	99.17	-	
3	-		-	
4	120	81.96	-	
5	-		-	
Monthly Bill (Monthly)	122.76	6756.40	36.830	2027.03
	Total(NRs)	7057.54		2777.03

Table 3: Comparison of Life Cycle Cost of FTL and LED light for 6 years of Operation

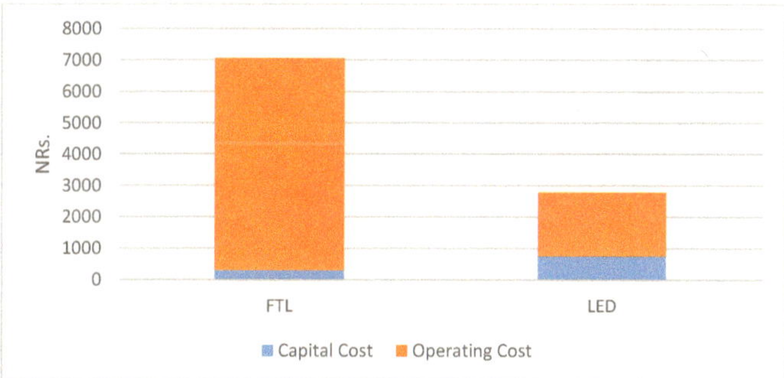

Figure 3: Comparison of Life Cycle Cost of FTL and LED light for 6 years of Operation

The detail calculation of costs involved in life cycle cost is given in ANNEX 5.

3.2 Economic Analysis with Solar Implementation

The stand-alone solar system is analyzed by software PVSYST and the cost of energy can be $ .31/kWh and the gross investment of $ 47892 for PV array size of 22.46 kWp. This is not viable cost economically and financially to implement this huge amount of solar system.

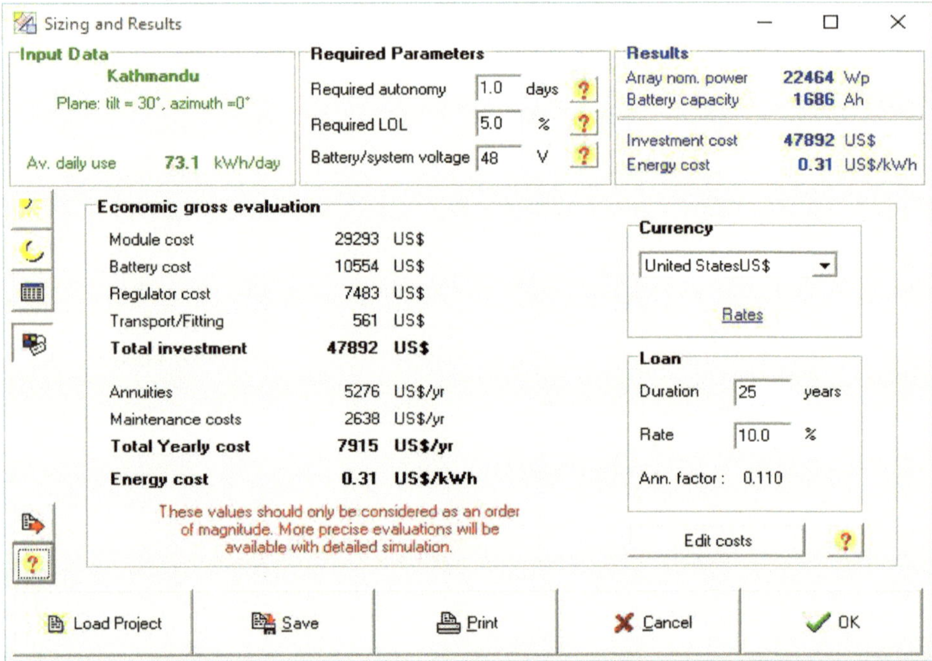

Figure 4: PVSYST window for costing solar system in existing condition

The stand-alone system for LED light based system is also analyzed and found to be energy cost of US $ 0.33/kWh and gross investment cost of $ 24633 for PV array size of 11.3 kWp.

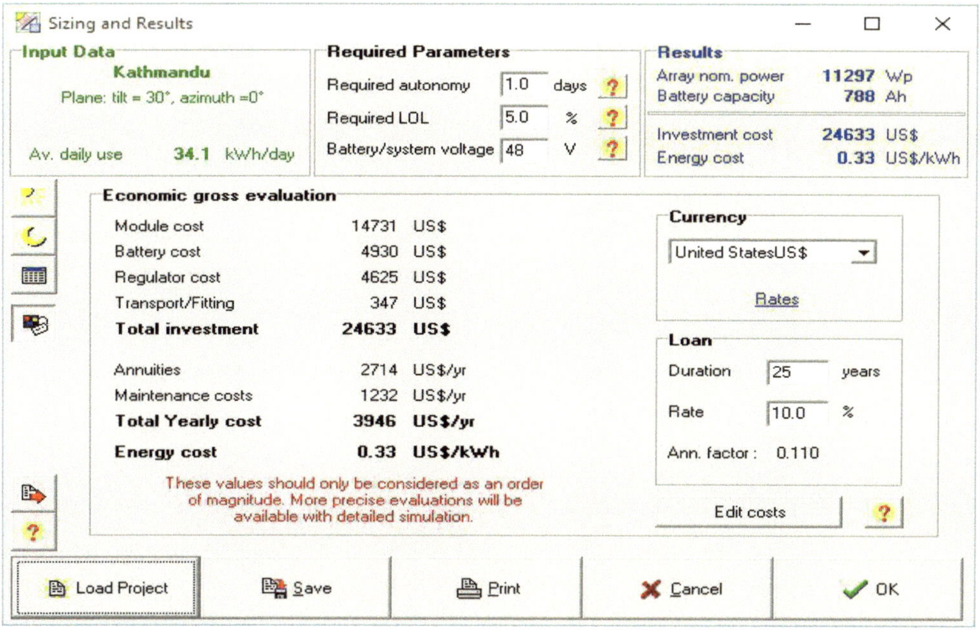

Figure 5: PVSYST window for costing of solar system for LED based system

The cost of investment for both options are very large in comparison to the energy bill of grid electricity. Also the cost of energy is also greater than the grid.

From these figures value, the implementation of solar system is not financially acceptable. But in Nepal, Due to load shedding, diesel is used to provide the reliable supply at huge cost of electricity, in average NRs.40/kWh. If this rate is considered for load shedding time, the solar system may be financially acceptable. The solar system is environmentally friendly and non-polluting source for power production, therefore, can be implemented in replacement of diesel generator set.

The quotation for 1kWp system is given in ANNEX 4. The PVSYST output of solar system in existing system and LED based system are given in ANNEX 6.

4. RISK ASSESSMENT

The implementation of project has certain financial and environmental risks which are discussed below.

4.1 Financial Risks

Financial risks are mainly involve with the implementation of solar system. While implementing the solar system for the diesel generator replacement, the financial risks are involved in the project life. The project life is 25 years and load shedding of Nepal could be made nearly zero for that period of time and within two years, load shedding is drastically reduced when Upper-Tamakoshi of capacity 456MW is completed by 2016. Thus, the cost of energy from grid is reliable and cheaper than the solar power system and it creates financial risks in implementation of solar PV system.

4.2 Environmental Risks

Environmental risks involves in both the options; solar PV system implementation and CFL replaced by LED light.

The main waste that will be produced by the project is used lead acid batteries, that require replacing every eight years or so. Solar panels have a guaranteed life of 25 years, therefore, they will not produce waste for a considerable period of time unless they are physically damaged.

Lead acid batteries need to be recycled to avoid pollution from the lead plates and the sulphuric acid that the plates sit in. Recycling involves the recovery of lead to produce new batteries. Local scrap dealers collect lead for (i) crushing then transportation to India for processing, or (ii) local recycling in small quantities by a battery company, however, neither method is environmentally friendly.

Thin-film PV cells contain a number of more toxic materials than those used in traditional silicon photovoltaic cells, including gallium arsenide, copper-indium-gallium-diselenide, and cadmium-telluride. If not handled and disposed of properly, these materials could pose serious environmental or public health threats. Other impact are related to the land use, visual impact, waste management etc.

When CFLs and FTLs are replaced by LED light, the CFLs and FTLs have no any use in any equipment and process or scrap value for a product and thus is disposed and may cause environmental and health hazards.

Each and every atom of mercury is capable bio-chemically to disable an enzyme or other critical protein in our body. Thus, mercury has the potential to produce significant health effects through series of bio chemical reactions in our body. The mercury ions can cause irreversible damages to several organs, when animals and human organisms, are exposed to them. Even worse, they could lead the organism to death. As well as the inhalation of mercury vapors, the ingestion of food containing organo-mercury, especially methyl mercury (CH3Hg+), represents great health risks.
Typically, a CFL contains 1 mg to 5mg of mercury and higher values for FTL. The estimated weight of a typical CFL is 0.04 kg without base, which is expected to contain around 20% of waste (by weight) as mercury-contaminated waste. The typical 1.2-metre fluorescent lamps contain approximately 0.26 kg of glass, 0.02 kg of combined metals and 0.01 kg of phosphor powder [6].

5. ENERGY SAVING STRATEGY MODELLING

After studying the energy consumption, expected area of saving and its economic analysis along with the risk assessment of the expected area of saving, the energy saving strategy is modelling in two ways:

5.1 CFLs and FTLs are replaced by LED light.

The CFLs and FTLs are not replaced at a single stage of replacement rather it is replaced in many different stages or milestone. These stages follow the following activity as a stage for replacement:

- The ageing effect of light also emits low light energy and these are replaced in one stage.
- When the CFLs or FTLs are burnt out or damaged or not working by any means and this condition is found frequently then it is replaced by LED light of corresponding equivalent. This activity is frequent and is also a stage.
- A periodic replacement plan is made for low grade luminaire with appropriate and equivalent LED light.

This stage plan is shown diagrammatically in figure and there is a tie interval between two stages.

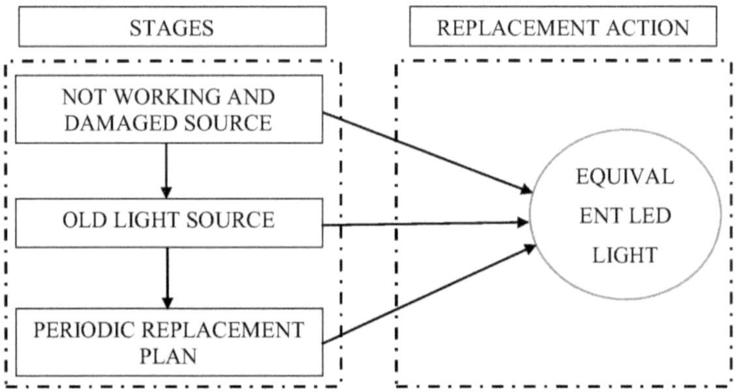

Figure 6: Model for replacement of CFLs and FTLs by LED light

This model also reduces high initial investment cost for replacement by LED light and also lower present worth because the investment is divided in certain time interval by creation of the stages in the model.

5.2 Solar-Grid Hybrid System

Nepal suffers an energy crises from several years and increasing demand of electricity increase the demand of diesel generator and whose energy cost of production is about NRs. 40/kWh. Also carbon trading and promotion for the use of renewable energy sources leads to solar PV system. The availability of sun light made solar PV system more feasible for wide range of geographic area. The stand-alone system has high cost of electricity generation and hence the hybrid system should be implemented.

For instance of that building, the 40% of power is provided by solar PV system and hence 4kWp solar PV system should be preferred. The cost of electric energy for this system is $0.18/kWh which is lower value than the stand-alone. The cost of electricity is calculated using PVSYST is shown in figure 7.

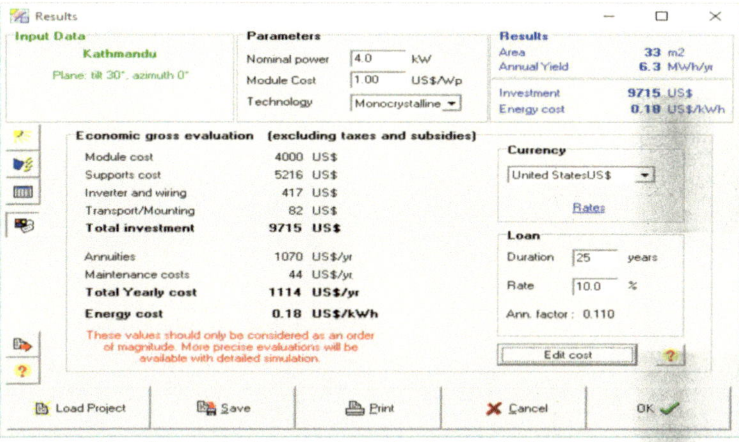

Figure 7: Cost of electricity generation for Hybrid system with 4kWp solar PV system

The PVSYST calculation sheet is in ANNEX 6. The model for this system is shown in figure 8 below.

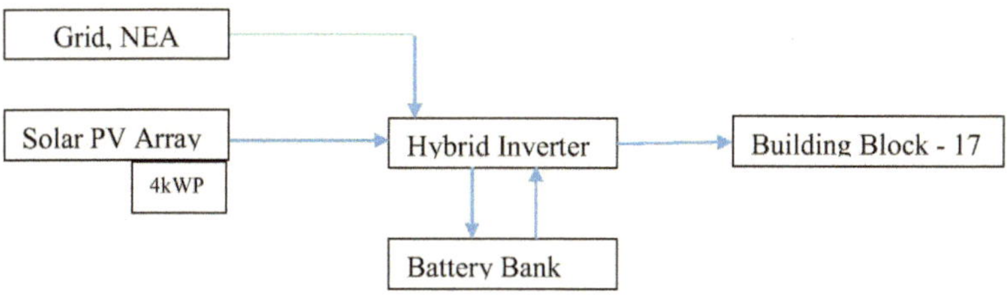

Figure 8: Model for Solar-Grid Hybrid System

6. CONCLUSION

The preliminary energy audit for girls' hostel (Block-17) shows different energy management process such as solar implementation for existing system and LED based system, LED based plan by replacing CFLs and FTLs. These energy management plans have its own environmental and financial risks and thus two different models are planned for energy management process.

The best plan from the two models either individually or in synergism go further for detail energy audit study based on advantages and dis-advantages that, under difference circumstances and from different perspectives. In conclusion of the preliminary energy audit, the synergism makes one more appealing than the others for implementation decision in girls' hostel which not only save the energy but also the cost of energy with reduction of maximum demand of energy.

REFERENCES

[1] IEA, "World Energy Outlook," International Energy Agency, 2010.

[2] WECS, "National Survey of Energy Consumption and Supply Situation in Nepal," Water and Energy Commision Secretariat, 2010.

[3] "Kathmandu University," 06 07 2014. [Online]. Available: http://www.ku.edu.np/university/index.php?go=physical#17. [Accessed 06 07 2015].

[4] OSRAM COMPANY, "OSRAM DULUX® EL ELECTRONIC ENERGY SAVING LAMPS," OSRAM WI 3/00 Co..

[5] Philips, [Online]. Available: www.philips.com/masterled. [Accessed 07 08 2015].

[6] "Guidelines for Environmentally Sound Mercury Management," Central Pollution Ccontrol Board, Ministry of Environment of Forests, Delhi, 2008.

ANNEX 1: Illuminance Level

S.N.	Room No.	Morning(Lux) 8:00 AM	Evening(Lux) 8:00 PM
1	101/Bathroom	35	55
2	102	87	88
3	103	120	76
4	104	390	86
5	105	200	44
6	106	193	76
7	107	68	77
8	108	68	88
9	109/Bathroom	-	55
10	110	265	81
11	111	46	55
12	201/Waiting Room	45	47
13	202/Reception	54	44
14	203/Store	35	55
15	204	383	67
16	205	483	89
17	206/Pantry	-	36
18	207	129	53
19	208/Bathroom	-	55
20	209	228	69
21	210	326	93
22	211/Bathroom	-	55
23	212	448	76
24	213/Common Room	130	50
25	214	191	78
26	215	385	60
27	216	213	76
28	217/Kitchen	-	44
29	218	391	57
30	219	278	91
31	220/Bathroom	-	55
32	221	246	59
33	222	385	61
34	223	401	45
35	224	342	41
36	225/Bathroom	-	85
37	226	356	77

38	227	151	48
39	228	117	90
40	301	331	56
41	302	132	34
42	303/Bathroom	-	55
43	304	77	65
44	305	79	74
45	306/Pantry	-	50
46	307	444	46
47	308/Bathroom	-	55
48	309	98	39
49	310	337	65
50	311/Bathroom	-	55
51	312	155	61
52	313/Common Room	130	56
53	314	345	54
54	315	343	56
55	316	234	76
56	317	321	55
57	318	332	44
58	319	293	37
59	320/Bathroom	-	55
60	321	322	34
61	322	246	56
62	323	119	47
63	324	123	62
64	325/Bathroom	-	55
65	326	85	93
66	327	347	65
67	328	337	48
68	401	238	44
69	402	150	64
70	403/Bathroom	-	55
71	404	56	77
72	405	250	86
73	406/Pantry	-	50
74	407	349	89
75	408/Bathroom	-	55
76	409	395	54
77	410	375	68
78	411/Bathroom	-	55
79	412	232	86

80	413/Common Room	130	44
81	414	239	57
82	415	394	42
83	416	201	66
84	417	339	68
85	418	257	87
86	419	244	65
87	420/Bathroom	-	96
88	421	176	55
89	422	146	75
90	423	345	54
91	424	429	84
92	425/Bathroom	-	55
93	426	226	88
94	427	192	45
95	428	384	54
96	501/Bathroom	-	55
97	502	173	96
98	503	149	30
99	504	193	44
100	505	175	57
101	506	178	70
102	507	183	75
103	508	348	87
104	509/Pantry	-	55
105	510	55	86
106	511	356	65
107	512	178	71
108	513	332	49
109	514	414	65
110	515/Bathroom	-	55
111	516	342	69
112	517	234	56

*Illuminance level is measured on 27 July, 2015

ANNEX 2: Room wise Individual consumption of CFLs and FTLs

S.N.	Room No.	Equipments	Quantity	Power (W)	Total Power (W)	Operating Hour (hr)	Total (Wh)
1	101	CFL	5	11	55	10	550
2	102	FTL	1	40	40	8	320
		CFL	2	11	22	5	110
3	103	FTL	1	40	40	8	320
		CFL	2	11	22	5	110
4	104	FTL	1	40	40	8	320
		CFL	2	11	22	5	110
5	105	FTL	1	40	40	8	320
		CFL	2	11	22	5	110
6	106	FTL	2	40	80	4	320
		CFL	3	11	33	5	165
7	107	FTL	1	40	40	8	320
		CFL	2	11	22	5	110
8	108	FTL	1	40	40	8	320
		CFL	2	11	22	5	110
9	109	CFL	5	11	55	10	550
10	110	FTL	1	40	40	8	320
		CFL	2	11	22	5	110
11	111	FTL	1	40	40	8	320
		CFL	2	11	22	5	110
12	201/Waiting Room	FTL	1	40	40	8	320
		CFL	4	11	44	5	220
13	202/Reception	FTL	1	40	40	8	320
		CFL	2	11	22	5	110
		LCD Computer	1	85	85	12	1020
		Printer	1	550	550	0	0
14	203/Store	CFL	2	18	36	2	72
15	204	FTL	1	40	40	8	320
		CFL	2	11	22	5	110
16	205	FTL	1	40	40	8	320
		CFL	2	11	22	5	110
17	206	CFL	5	18	90	10	900
18	207	FTL	1	40	40	8	320
		CFL	2	11	22	5	110
19	208	CFL	5	18	90	4	360
20	209	FTL	1	40	40	8	320
		CFL	2	11	22	5	110
21	210	FTL	1	40	40	8	320

		CFL	2	11	22	5	110
22	211	CFL	5	18	90	10	900
23	212	FTL	1	40	40	8	320
		CFL	2	11	22	5	110
24	213	FTL	1	40	40	8	320
		LCD TV	1	40	40	4	160
		CFL	2	11	22	5	110
25	214	FTL	1	40	40	8	320
		CFL	2	11	22	5	110
26	215	FTL	1	40	40	8	320
		CFL	2	11	22	5	110
27	216	FTL	1	40	40	8	320
		CFL	2	11	22	5	110
28	217	CFL	2	12	24	4	96
		FTL	1	40	40	9	360
		Rice Cooker	1	1000	1000	1	1000
		Fridge	1	500	500	12	6000
		Heater	1	1200	1200	4	4800
		Iron	1	1000	1000	1	1000
29	218	FTL	1	40	40	8	320
		CFL	2	11	22	5	110
30	219	FTL	1	40	40	8	320
		CFL	2	11	22	5	110
31	220	CFL	5	18	90	4	360
32	221	FTL	1	40	40	8	320
		CFL	2	11	22	5	110
33	222	FTL	1	40	40	8	320
		CFL	2	11	22	5	110
34	223	FTL	1	40	40	8	320
		CFL	2	11	22	5	110
35	224	FTL	1	40	40	8	320
		CFL	2	11	22	5	110
36	225	CFL	5	18	90	10	900
37	226	FTL	1	40	40	8	320
		CFL	2	11	22	5	110
38	227	FTL	1	40	40	8	320
		CFL	2	11	22	5	110
39	228	FTL	1	40	40	8	320
		CFL	2	11	22	5	110
40	301	FTL	1	40	40	8	320
		CFL	4	11	44	5	220
41	302	FTL	1	40	40	8	320

		CFL	2	11	22	5	110
42	303	CFL	2	18	36	2	72
43	304	FTL	1	40	40	8	320
		CFL	2	11	22	5	110
44	305	FTL	1	40	40	8	320
		CFL	2	11	22	5	110
45	306	CFL	5	18	90	10	900
46	307	FTL	1	40	40	8	320
		CFL	2	11	22	5	110
47	308	CFL	5	18	90	4	360
48	309	FTL	1	40	40	8	320
		CFL	2	11	22	5	110
49	310	FTL	1	40	40	8	320
		CFL	2	11	22	5	110
50	311	CFL	5	18	90	10	900
51	312	FTL	1	40	40	8	320
		CFL	2	11	22	5	110
52	313	FTL	1	40	40	8	320
		LCD TV	1	40	40	4	160
		CFL	2	11	22	5	110
53	314	FTL	1	40	40	8	320
		CFL	2	11	22	5	110
54	315	FTL	1	40	40	8	320
		CFL	2	11	22	5	110
55	316	FTL	1	40	40	8	320
		CFL	2	11	22	5	110
56	317	CFL	2	12	24	4	96
		FTL	1	40	40	9	360
57	318	FTL	1	40	40	8	320
		CFL	2	11	22	5	110
58	319	FTL	1	40	40	8	320
		CFL	2	11	22	5	110
59	320	CFL	5	18	90	4	360
60	321	FTL	1	40	40	8	320
		CFL	2	11	22	5	110
61	322	FTL	1	40	40	8	320
		CFL	2	11	22	5	110
62	323	FTL	1	40	40	8	320
		CFL	2	11	22	5	110
63	324	FTL	1	40	40	8	320
		CFL	2	11	22	5	110
64	325	CFL	5	18	90	10	900

65	326	FTL	1	40	40	8	320
		CFL	2	11	22	5	110
66	327	FTL	1	40	40	8	320
		CFL	2	11	22	5	110
67	328	FTL	1	40	40	8	320
		CFL	2	11	22	5	110
68	401	FTL	1	40	40	8	320
		CFL	4	11	44	5	220
69	402	FTL	1	40	40	8	320
		CFL	2	11	22	5	110
70	403	CFL	2	18	36	2	72
71	404	FTL	1	40	40	8	320
		CFL	2	11	22	5	110
72	405	FTL	1	40	40	8	320
		CFL	2	11	22	5	110
73	406	CFL	5	18	90	10	900
74	407	FTL	1	40	40	8	320
		CFL	2	11	22	5	110
75	408	CFL	5	18	90	4	360
76	409	FTL	1	40	40	8	320
		CFL	2	11	22	5	110
77	410	FTL	1	40	40	8	320
		CFL	2	11	22	5	110
78	411	CFL	5	18	90	10	900
79	412	FTL	1	40	40	8	320
		CFL	2	11	22	5	110
80	413	FTL	1	40	40	8	320
		LCD TV	1	40	40	4	160
		CFL	2	11	22	5	110
81	414	FTL	1	40	40	8	320
		CFL	2	11	22	5	110
82	415	FTL	1	40	40	8	320
		CFL	2	11	22	5	110
83	416	FTL	1	40	40	8	320
		CFL	2	11	22	5	110
84	417	CFL	2	12	24	4	96
		FTL	1	40	40	9	360
85	418	FTL	1	40	40	8	320
		CFL	2	11	22	5	110
86	419	FTL	1	40	40	8	320
		CFL	2	11	22	5	110
87	420	CFL	5	18	90	4	360

88	421	FTL	1	40	40	8	320
		CFL	2	11	22	5	110
89	422	FTL	1	40	40	8	320
		CFL	2	11	22	5	110
90	423	FTL	1	40	40	8	320
		CFL	2	11	22	5	110
91	424	FTL	1	40	40	8	320
		CFL	2	11	22	5	110
92	425	CFL	5	18	90	10	900
93	426	FTL	1	40	40	8	320
		CFL	2	11	22	5	110
94	427	FTL	1	40	40	8	320
		CFL	2	11	22	5	110
95	428	FTL	1	40	40	8	320
		CFL	2	11	22	5	110
96	501	CFL	5	11	55	10	550
97	502	FTL	1	40	40	8	320
		CFL	2	11	22	5	110
98	503	FTL	1	40	40	8	320
		CFL	2	11	22	5	110
99	504	FTL	1	40	40	8	320
		CFL	2	11	22	5	110
100	505	FTL	1	40	40	8	320
		CFL	2	11	22	5	110
101	506	FTL	2	40	80	4	320
		CFL	3	11	33	5	165
102	507	FTL	1	40	40	8	320
		CFL	2	11	22	5	110
103	508	FTL	1	40	40	8	320
		CFL	2	11	22	5	110
104	509	CFL	5	11	55	10	550
105	510	FTL	1	40	40	8	320
		CFL	2	11	22	5	110
106	511	FTL	1	40	40	8	320
		CFL	2	11	22	5	110
107	512	FTL	2	40	80	4	320
		CFL	3	11	33	5	165
108	513	FTL	1	40	40	8	320
		CFL	2	11	22	5	110
109	514	FTL	1	40	40	8	320
		CFL	2	11	22	5	110
110	515	CFL	5	11	55	10	550

111	516	FTL	1	40	40	8	320
		CFL	2	11	22	5	110
112	517	FTL	1	40	40	8	320
		CFL	2	11	22	5	110
	Corridor 1	CFL	16	11	176	12	2112
	Corridor 2	CFL	42	11	462	12	5544
	Corridor 3	CFL	42	11	462	12	5544
	Corridor 4	CFL	42	11	462	12	5544
	Corridor 5	CFL	17	11	187	12	2244
		Total			**13680**		**87357**

ANNEX 3: Room wise Individual consumption of LED (when replaced)

S.N.	Room No.	Equipments	Quantity	Power (W)	Total Power (W)	Opearting Hour (hr)	Total (Wh)
1	101	LED	5	7	35	10	350
2	102	LED	1	12	12	8	96
		LED	2	7	14	5	70
3	103	LED	1	12	12	8	96
		LED	2	7	14	5	70
4	104	LED	1	12	12	8	96
		LED	2	7	14	5	70
5	105	LED	1	12	12	8	96
		LED	2	7	14	5	70
6	106	LED	2	12	24	4	96
		LED	3	7	21	5	105
7	107	LED	1	12	12	8	96
		LED	2	7	14	5	70
8	108	LED	1	12	12	8	96
		LED	2	7	14	5	70
9	109	LED	5	7	35	10	350
10	110	LED	1	12	12	8	96
		LED	2	7	14	5	70
11	111	LED	1	12	12	8	96
		LED	2	7	14	5	70
12	201/Waiting Room	LED	1	12	12	8	96
		LED	4	7	28	5	140
13	202/Reception	LED	1	12	12	8	96
		LED	2	7	14	5	70
		LCD Computer	1	0	0	12	0
		Printer	1	0	0	0	0
14	203/Store	LED	2	7	14	2	28
15	204	LED	1	12	12	8	96
		LED	2	7	14	5	70
16	205	LED	1	12	12	8	96
		LED	2	7	14	5	70
17	206	LED	5	7	35	10	350
18	207	LED	1	12	12	8	96
		LED	2	7	14	5	70
19	208	LED	5	7	35	4	140
20	209	LED	1	12	12	8	96

		LED	2	7	14	5	70
21	210	LED	1	12	12	8	96
		LED	2	7	14	5	70
22	211	LED	5	7	35	10	350
23	212	LED	1	12	12	8	96
		LED	2	7	14	5	70
24	213	LED	1	12	12	8	96
		LCD TV	1	0	0	4	0
		LED	2	7	14	5	70
25	214	LED	1	12	12	8	96
		LED	2	7	14	5	70
26	215	LED	1	12	12	8	96
		LED	2	7	14	5	70
27	216	LED	1	12	12	8	96
		LED	2	7	14	5	70
28	217	LED	2	7	14	4	56
		LED	1	12	12	9	108
		Rice Cooker	1	0	0	1	0
		Fridge	1	0	0	12	0
		Heater	1	0	0	4	0
		Iron	1	0	0	1	0
29	218	LED	1	12	12	8	96
		LED	2	7	14	5	70
30	219	LED	1	12	12	8	96
		LED	2	7	14	5	70
31	220	LED	5	7	35	4	140
32	221	LED	1	12	12	8	96
		LED	2	7	14	5	70
33	222	LED	1	12	12	8	96
		LED	2	7	14	5	70
34	223	LED	1	12	12	8	96
		LED	2	7	14	5	70
35	224	LED	1	12	12	8	96
		LED	2	7	14	5	70
36	225	LED	5	7	35	10	350
37	226	LED	1	12	12	8	96
		LED	2	7	14	5	70
38	227	LED	1	12	12	8	96
		LED	2	7	14	5	70
39	228	LED	1	12	12	8	96
		LED	2	7	14	5	70
40	301	LED	1	12	12	8	96

		LED	4	7	28	5	140
41	302	LED	1	12	12	8	96
		LED	2	7	14	5	70
42	303	LED	2	7	14	2	28
43	304	LED	1	12	12	8	96
		LED	2	7	14	5	70
44	305	LED	1	12	12	8	96
		LED	2	7	14	5	70
45	306	LED	5	7	35	10	350
46	307	LED	1	12	12	8	96
		LED	2	7	14	5	70
47	308	LED	5	7	35	4	140
48	309	LED	1	12	12	8	96
		LED	2	7	14	5	70
49	310	LED	1	12	12	8	96
		LED	2	7	14	5	70
50	311	LED	5	7	35	10	350
51	312	LED	1	12	12	8	96
		LED	2	7	14	5	70
52	313	LED	1	12	12	8	96
		LCD TV	1	0	0	4	0
		LED	2	7	14	5	70
53	314	LED	1	12	12	8	96
		LED	2	7	14	5	70
54	315	LED	1	12	12	8	96
		LED	2	7	14	5	70
55	316	LED	1	12	12	8	96
		LED	2	7	14	5	70
56	317	LED	2	7	14	4	56
		LED	1	12	12	9	108
57	318	LED	1	12	12	8	96
		LED	2	7	14	5	70
58	319	LED	1	12	12	8	96
		LED	2	7	14	5	70
59	320	LED	5	7	35	4	140
60	321	LED	1	12	12	8	96
		LED	2	7	14	5	70
61	322	LED	1	12	12	8	96
		LED	2	7	14	5	70
62	323	LED	1	12	12	8	96
		LED	2	7	14	5	70
63	324	LED	1	12	12	8	96

		LED	2	7	14	5	70
64	325	LED	5	7	35	10	350
65	326	LED	1	12	12	8	96
		LED	2	7	14	5	70
66	327	LED	1	12	12	8	96
		LED	2	7	14	5	70
67	328	LED	1	12	12	8	96
		LED	2	7	14	5	70
68	401	LED	1	12	12	8	96
		LED	4	7	28	5	140
69	402	LED	1	12	12	8	96
		LED	2	7	14	5	70
70	403	LED	2	7	14	2	28
71	404	LED	1	12	12	8	96
		LED	2	7	14	5	70
72	405	LED	1	12	12	8	96
		LED	2	7	14	5	70
73	406	LED	5	7	35	10	350
74	407	LED	1	12	12	8	96
		LED	2	7	14	5	70
75	408	LED	5	7	35	4	140
76	409	LED	1	12	12	8	96
		LED	2	7	14	5	70
77	410	LED	1	12	12	8	96
		LED	2	7	14	5	70
78	411	LED	5	7	35	10	350
79	412	LED	1	12	12	8	96
		LED	2	7	14	5	70
80	413	LED	1	12	12	8	96
		LCD TV	1	0	0	4	0
		LED	2	7	14	5	70
81	414	LED	1	12	12	8	96
		LED	2	7	14	5	70
82	415	LED	1	12	12	8	96
		LED	2	7	14	5	70
83	416	LED	1	12	12	8	96
		LED	2	7	14	5	70
84	417	LED	2	7	14	4	56
		LED	1	12	12	9	108
85	418	LED	1	12	12	8	96
		LED	2	7	14	5	70
86	419	LED	1	12	12	8	96

		LED	2	7	14	5	70
87	420	LED	5	7	35	4	140
88	421	LED	1	12	12	8	96
		LED	2	7	14	5	70
89	422	LED	1	12	12	8	96
		LED	2	7	14	5	70
90	423	LED	1	12	12	8	96
		LED	2	7	14	5	70
91	424	LED	1	12	12	8	96
		LED	2	7	14	5	70
92	425	LED	5	7	35	10	350
93	426	LED	1	12	12	8	96
		LED	2	7	14	5	70
94	427	LED	1	12	12	8	96
		LED	2	7	14	5	70
95	428	LED	1	12	12	8	96
		LED	2	7	14	5	70
96	501	LED	5	7	35	10	350
97	502	LED	1	12	12	8	96
		LED	2	7	14	5	70
98	503	LED	1	12	12	8	96
		LED	2	7	14	5	70
99	504	LED	1	12	12	8	96
		LED	2	7	14	5	70
100	505	LED	1	12	12	8	96
		LED	2	7	14	5	70
101	506	LED	2	12	24	4	96
		LED	3	7	21	5	105
102	507	LED	1	12	12	8	96
		LED	2	7	14	5	70
103	508	LED	1	12	12	8	96
		LED	2	7	14	5	70
104	509	LED	5	7	35	10	350
105	510	LED	1	12	12	8	96
		LED	2	7	14	5	70
106	511	LED	1	12	12	8	96
		LED	2	7	14	5	70
107	512	LED	2	12	24	4	96
		LED	3	7	21	5	105
108	513	LED	1	12	12	8	96
		LED	2	7	14	5	70
109	514	LED	1	12	12	8	96

		LED	2	7	14	5	70
110	515	LED	5	7	35	10	350
111	516	LED	1	12	12	8	96
		LED	2	7	14	5	70
112	517	LED	1	12	12	8	96
		LED	2	7	14	5	70
	Corridor 1	LED	16	7	112	12	1344
	Corridor 2	LED	42	7	294	12	3528
	Corridor 3	LED	42	7	294	12	3528
	Corridor 4	LED	42	7	294	12	3528
	Corridor 5	LED	17	7	119	12	1428
		Total			**4268**		**34263**

ANNEX 4: Quotation for Solar System and LED Bulb

[removed for publication]

ANNEX 5: Calculation for Comparison OF LED and CFL Energy Cost

		LED		FTL	
Time (Month)	Discount Factor (for i=10%)	LED Bill (monthly)	Present Value	FTL Bill (monthly)	
0	1	36.83	36.83	122.76	122.76
1	0.991768	36.83	36.52683	122.76	121.7495
2	0.983604	36.83	36.22615	122.76	120.7473
3	0.975508	36.83	35.92795	122.76	119.7533
4	0.967478	36.83	35.6322	122.76	118.7676
5	0.959514	36.83	35.33889	122.76	117.7899
6	0.951615	36.83	35.04799	122.76	116.8203
7	0.943782	36.83	34.75949	122.76	115.8587
8	0.936013	36.83	34.47336	122.76	114.905
9	0.928308	36.83	34.18958	122.76	113.9591
10	0.920666	36.83	33.90815	122.76	113.021
11	0.913088	36.83	33.62903	122.76	112.0907
12	0.905572	36.83	33.3522	122.76	111.168
13	0.898117	36.83	33.07766	122.76	110.2529
14	0.890724	36.83	32.80537	122.76	109.3453
15	0.883392	36.83	32.53533	122.76	108.4452
16	0.87612	36.83	32.26751	122.76	107.5525
17	0.868908	36.83	32.00189	122.76	106.6672
18	0.861756	36.83	31.73846	122.76	105.7891
19	0.854662	36.83	31.4772	122.76	104.9183
20	0.847627	36.83	31.21809	122.76	104.0547
21	0.840649	36.83	30.96112	122.76	103.1981
22	0.833729	36.83	30.70625	122.76	102.3486
23	0.826866	36.83	30.45349	122.76	101.5061
24	0.82006	36.83	30.20281	122.76	100.6706
25	0.813309	36.83	29.95419	122.76	99.84187
26	0.806615	36.83	29.70761	122.76	99.02
27	0.799975	36.83	29.46307	122.76	98.2049
28	0.79339	36.83	29.22054	122.76	97.39651
29	0.786859	36.83	28.98001	122.76	96.59477
30	0.780382	36.83	28.74145	122.76	95.79964
31	0.773958	36.83	28.50486	122.76	95.01105
32	0.767587	36.83	28.27022	122.76	94.22895
33	0.761268	36.83	28.03751	122.76	93.45328

34	0.755002	36.83	27.80671	122.76	92.68401
35	0.748787	36.83	27.57782	122.76	91.92106
36	0.742623	36.83	27.3508	122.76	91.1644
37	0.73651	36.83	27.12566	122.76	90.41396
38	0.730447	36.83	26.90237	122.76	89.6697
39	0.724434	36.83	26.68092	122.76	88.93157
40	0.718471	36.83	26.46129	122.76	88.19951
41	0.712557	36.83	26.24347	122.76	87.47348
42	0.706691	36.83	26.02744	122.76	86.75343
43	0.700874	36.83	25.81319	122.76	86.0393
44	0.695105	36.83	25.60071	122.76	85.33106
45	0.689383	36.83	25.38997	122.76	84.62864
46	0.683708	36.83	25.18097	122.76	83.932
47	0.67808	36.83	24.97369	122.76	83.2411
48	0.672498	36.83	24.76811	122.76	82.55589
49	0.666962	36.83	24.56423	122.76	81.87631
50	0.661472	36.83	24.36202	122.76	81.20234
51	0.656027	36.83	24.16148	122.76	80.5339
52	0.650627	36.83	23.96259	122.76	79.87097
53	0.645271	36.83	23.76534	122.76	79.2135
54	0.63996	36.83	23.56971	122.76	78.56144
55	0.634692	36.83	23.37569	122.76	77.91475
56	0.629467	36.83	23.18327	122.76	77.27338
57	0.624286	36.83	22.99244	122.76	76.63729
58	0.619147	36.83	22.80317	122.76	76.00644
59	0.61405	36.83	22.61546	122.76	75.38078
60	0.608995	36.83	22.4293	122.76	74.76027
61	0.603982	36.83	22.24467	122.76	74.14487
62	0.59901	36.83	22.06156	122.76	73.53453
63	0.59408	36.83	21.87995	122.76	72.92922
64	0.589189	36.83	21.69984	122.76	72.32889
65	0.584339	36.83	21.52122	122.76	71.7335
66	0.579529	36.83	21.34406	122.76	71.14301
67	0.574759	36.83	21.16837	122.76	70.55739
68	0.570028	36.83	20.99411	122.76	69.97658
69	0.565335	36.83	20.8213	122.76	69.40056
70	0.560682	36.83	20.6499	122.76	68.82927
71	0.556066	36.83	20.47992	122.76	68.26269
72	0.551489	36.83	20.31134	122.76	67.70078
	Total		2027.031		6756.402

ANNEX 6: PVSYST Output

1. SOLAR SYSTEM FOR EXISTING SYSTEM

PVSYST V5.59	Ajay &Singh	07/08/15

Stand-alone system presizing

Geographical Site	**Kathmandu**		**Country**	**Nepal**
Situation	Latitude	27.5°N	Longitude	85.0°E
Time defined as	Solar Time		Altitude	1478 m
Collector Plane Orientation	Tilt	30°	Azimuth	0°

System pre-sizing evaluation

Average use of energy	Daily	73.1 kWh/day	Yearly	26679 kWh
Autonomy		1.0 days		
Loss-of-Load	Time fraction	5.0 %	Missing energy	1261 kWh
Battery system	Voltage	48 V	Capacity	1686 Ah
PV array	Nominal power	22.58 kWp	Nominal Current	362 A
Economic gross evaluation	Investment	48065 US$	Energy price	0.31 US$/kWh

PV energy yield and user's needs

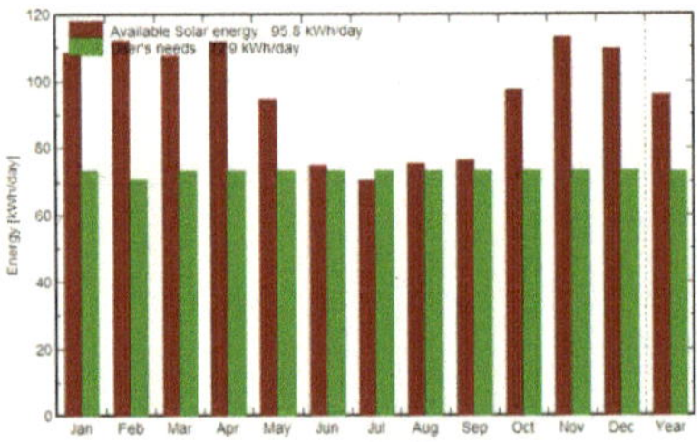

Battery SOC and Loss-of-Load Probability

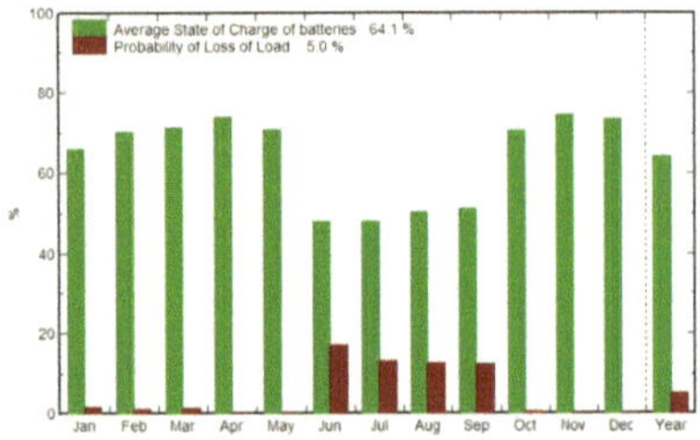

	Incld.	PV avail.	Demand	Excess	Missing	SOC	Pr. LOL	Fuel
	kWh/m².day	kWh	kWh	kWh	kWh	%	%	liter
Jan.	6.0	3362.7	2265.9	849.7	35.5	66	1.6	23.7
Feb.	6.4	3249.4	2046.6	1014.0	25.5	70	1.2	17.0
Mar.	6.0	3336.6	2265.9	797.2	33.7	71	1.5	22.5
Apr.	6.2	3348.7	2192.8	891.5	0.0	74	0.0	0.0
May	5.2	2928.4	2265.9	458.0	0.0	71	0.0	0.0
June	4.1	2238.0	2192.8	115.2	346.5	48	17.0	231.0
July	3.9	2179.4	2265.9	41.5	282.5	48	13.0	188.3
Aug.	4.2	2335.6	2265.9	107.5	266.0	50	12.5	177.3
Sep.	4.2	2292.7	2192.8	63.9	255.8	51	12.4	170.5
Oct.	5.4	3019.9	2265.9	511.5	15.5	71	0.7	10.3
Nov.	6.3	3387.9	2192.8	901.1	0.0	74	0.0	0.0
Dec.	6.1	3391.3	2265.9	835.1	0.0	73	0.0	0.0
Year	5.3	35070.	26678.	6586.3	1260.9	64	5.0	840.6

2. SOLAR SYSTEM FOR LED BASED SYSTEM

PVSYST V5.59	Ajay &Singh	07/08/15

SOLAR SYSTEM SIZING FOR LED BASED SYSTEM

Stand-alone system presizing

Geographical Site	**Kathmandu**		**Country**	**Nepal**
Situation	Latitude	27.5°N	Longitude	85.0°E
Time defined as	Solar Time		Altitude	1478 m
Collector Plane Orientation	Tilt	30°	Azimuth	0°

System pre-sizing evaluation

Average use of energy	Daily	34.1 kWh/day	Yearly	12463 kWh
Autonomy		1.0 days		
Loss-of-Load	Time fraction	5.0 %	Missing energy	584 kWh
Battery system	Voltage	48 V	Capacity	788 Ah
PV array	Nominal power	11.30 kWp	Nominal Current	181 A
Economic gross evaluation	Investment	24633 US$	Energy price	0.33 US$/kWh

PV energy yield and user's needs

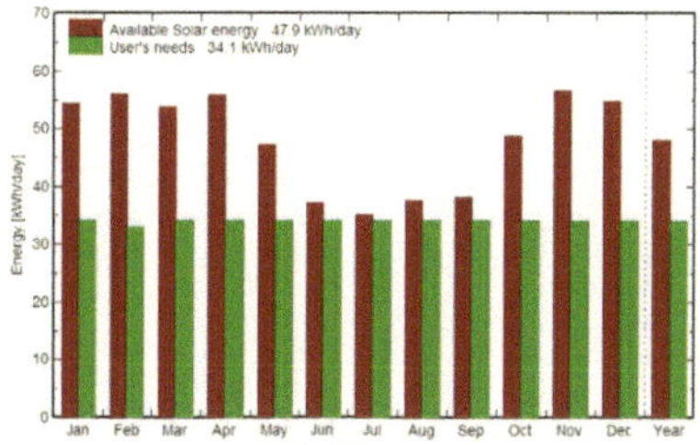

Battery SOC and Loss-of-Load Probability

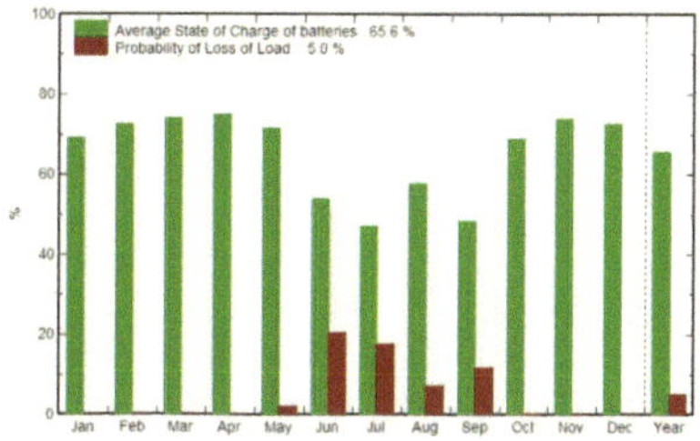

	Incid.	PV avail.	Demand	Excess	Missing	SOC	Pr. LOL	Fuel
	kWh/m².day	kWh	kWh	kWh	kWh	%	%	liter
Jan.	6.0	1682.7	1058.5	489.3	0.0	69	0.0	0.0
Feb.	6.4	1626.0	956.0	550.7	0.0	73	0.0	0.0
Mar.	6.0	1669.7	1058.5	482.7	0.0	74	0.0	0.0
Apr.	6.2	1675.7	1024.3	527.9	0.0	75	0.0	0.0
May	5.2	1465.4	1058.5	306.2	23.1	71	2.2	15.4
June	4.1	1119.9	1024.3	223.4	187.4	54	20.6	124.9
July	3.9	1090.6	1058.5	82.1	180.7	47	17.5	120.5
Aug.	4.2	1168.7	1058.5	77.2	66.6	58	7.2	44.4
Sep.	4.2	1147.3	1024.3	126.4	121.8	48	11.9	81.2
Oct.	5.4	1511.2	1058.5	328.5	4.9	69	0.5	3.2
Nov.	6.3	1695.3	1024.3	541.8	0.0	74	0.0	0.0
Dec.	6.1	1697.0	1058.5	502.9	0.0	73	0.0	0.0
Year	5.3	17549.	12462.	4239.1	584.4	66	5.0	389.6

3. SOLAR SYSTEM FOR HYBRID SYSTEM

PVSYST V5.59	Ajay &Singh	08/08/15

HYBID SYSTEM SIZING

Grid system presizing

Geographical Site	**Kathmandu**		**Country**	**Nepal**
Situation	Latitude	27.5°N	Longitude	85.0°E
Time defined as	Solar Time		Altitude	1478 m
Collector Plane Orientation	Tilt	30°	Azimuth	0°

PV-field installation main features

Module type	Standard
Technology	Monocrystalline cells
Mounting method	Facade or tilt roof
Back ventilation properties	Ventilated

System characteristics and pre-sizing evaluation

PV-field nominal power (STC)	Pnom	4.0 kWp		
Collector area	Acoll	33 m²		
Annual energy yield	Eyear	6.26 MWh	Specific yield	1564 kWh/kWp
Economic gross evaluation	Investment	9715 US$	Energy price	0.18 US$/kWh

Meteo and incident energy

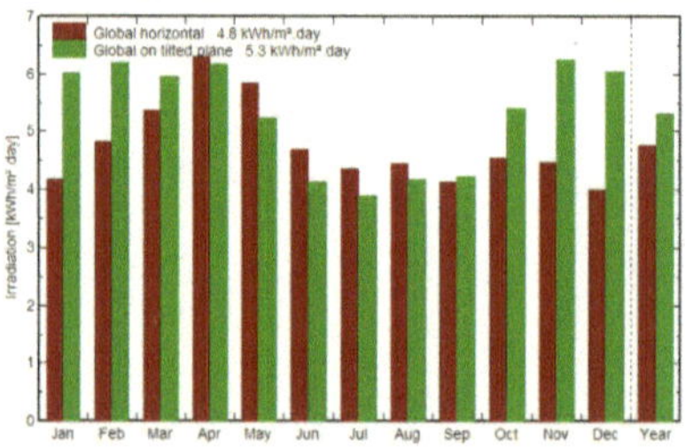

System output

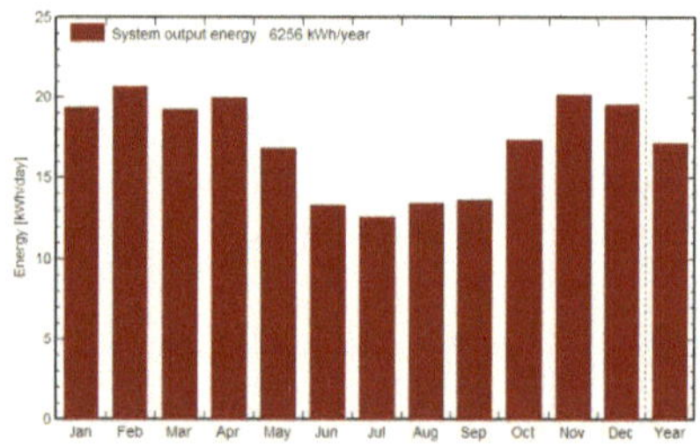

	Gl. horiz.	Coll. Plane	System output	System output
	kWh/m².day	kWh/m².day	kWh/day	kWh
Jan.	4.16	6.01	19.35	600
Feb.	5.00	6.43	20.70	580
Mar.	5.35	5.96	19.20	595
Apr.	6.30	6.18	19.91	597
May	5.83	5.23	16.85	522
June	4.69	4.13	13.31	399
July	4.35	3.89	12.54	389
Aug.	4.44	4.17	13.44	417
Sep.	4.13	4.23	13.63	409
Oct.	4.55	5.39	17.38	539
Nov	4.47	6.25	20.14	604
Dec.	4.00	6.06	19.51	605
Year	4.77	5.32	17.14	6256